German Heavy Half-Prime Movers 1934-1945

by Reinhard Frank

Translated from the German by Don Cox

The German half-tracked prime movers at a glance (from left): lt. half-track (Sd. Kfz 10) prototype; 3-ton lt. half-track (Sd. Kfz 11); 5-ton med. half-track (Sd. Kfz. 6); 8-ton med. half-track (Sd. Kfz. 7); 12-ton hvy. half-track (Sd. Kfz. 8) 18-ton hvy. half-track (Sd. Kfz. 9). Notice the uncommon four-row seating design and the spoked wheels of the 18-tonner.

Schiffer Military/Aviation History
Atglen, PA

Acknowlegements

My particular thanks go to Gertrud, Stefan and Daniel. Valuable photos and documents were provided by: Horst Beiersdorf of the Bundesarchiv Koblenz, Siegfried Bunke, Siegfried Ehrt, Michael Foedrowitz, Henry Hoppe, Hansludwig Huber, Randolf Kugler, Horst Scheibert, Franz Sindler, Peter Taghon, August Welte and many others.

The cover picture was painted by Heinz Rode; the drawings on the inner cover were made available by Horst Hellman.

Sources:
- Company documents
- Original regulations
- Archival material
- Original photos taken by combatants during the war.

The Geb. Art. Rgt. 82 must certainly have taken the prize for the Wehrmacht's "best camouflage cover" for this carefully concealed prime mover and gun. The "sandal" insignia of the 7th Gebirgsdivision can just be made out on the vehicle's fender.

Translated from the German by Don Cox

Copyright © 1996 by Schiffer Publishing, Ltd.

All rights reserved. No part of this work may be reproduced or used in any forms or by any means—graphic, electronic or mechanical, including photocopying or information storage and retrieval systems—without written permission from the copyright holder.

Printed in China.
ISBN: 0-7643-0167-5

This book was originally published under the title,
Waffen Arsenal-Die Schweren Zugkraftwagen Der Wehrmacht 1934-1945
by Podzun-Pallas Verlag

We are interested in hearing from authors with book ideas on related topics.

Published by Schiffer Publishing Ltd.
77 Lower Valley Road
Atglen, PA 19310
Please write for a free catalog.
This book may be purchased from the publisher.
Please include $2.95 postage.
Try your bookstore first.

Cover Picture:
18-ton heavy half-tracked prime mover (Sd. Kfz. 9)

THE Sd. Kfz. 8 12-TON HEAVY HALF-TRACKED PRIME MOVER

The 12-tonner was, according to D 608, "designed for pulling trailer loads." On the basis of the "Classified Data Sheets for Army Weapons, Vehicles and Equipment", this included the 15 cm Kanone 16 (multi- and single-piece towed loads), the 15 cm Kanone 18 (multi- and single-piece towed loads), the 10.5 cm Flak and the 21 cm Mörser.

Primary contractor for the 12-tonner was Daimler-Benz AG. Approximately 4000 of the type were built at the following sites: Daimler-Benz, Werk 40, Berlin-Marienfelde; Krauss-Maffei, Munich-Allach; Friedrich Krupp AG, Essen and Mühlhausen (Alsace). Interestingly, after the war the Czech army continued to use the 12-tonner. These vehicles were built indigenously, with parts apparently coming from so-called "relocated operations."

Daimler-Benz gathered its first experience with half-tracked vehicles as early as 1931 with the development of the Type ZD 5. This was a rather unattractive-looking half-track design powered by a 150 hp 12-cylinder Maybach Type DSO 8 engine. In order to circumvent the Versailles Treaty development and testing was undertaken in a cooperative effort with the then-friendly Soviet Union.

In 1934 Daimler-Benz produced what was at the time the heaviest half-track of the Wehrmacht, the "heavy all-terrain half-tracked prime mover (Sd. Kfz. 8), designated "DBs 7 Baujahr 1934." The vehicle was driven by the same engine as the ZD 5. According to D 601 from 8 Nov 1935 its tow strength was listed as 8 tons, although this was meant to reflect the pulling power and not the actual maximum weight of the towed load.

In 1936 the model DBs 8 (also written as DB s 8) appeared; this was the final design of the 12-tonner series and had the standard form of all German half-tracks. As with its predecessors, the vehicle was also powered by the Type DSO 8 Maybach engine. Towed load was given as 12 tons and it was designated as "schwerer Zugkraftwagen 12 t (Sd. Kfz. 8), or "12-ton heavy half-tracked prime mover."

The DB 9 version (the "s" was dropped in the company's designation) was built in the years 1938 and 1939 and was now fitted with the 12-cylinder Maybach Type HL 85 TUKRM engine having an output of 185 hp.

The Type DB 10, built from 1939 to 1944, differed from previous models primarily through the use of a torsion bar suspension in place of the former leaf spring suspension. Towed load ("weight, normal") was given as 14 tons for the DB 10 Baujahr 1939/40 in D 608/11 from 12 Jan 1940; however the "Classified Data Sheets" the load remained at 12 tons.

The 12-tonner was a rugged and reliable vehicle — but because of economic cut-backs continued production would have been impractical following the introduction of the 18-tonner.

According to the "Classified Data Sheets", one half-track cost 46000 RM and reputedly took an average of 15 months to build.

TECHNICAL DATA BASED ON D 608/1 FOR THE 12-TON HEAVY HALF-TRACKED PRIME MOVER (Sd. Kfz. 8), TYPE DB 10 BAUJAHR 1939/40:

Vehicle Performance

Towed load (weight) normal	14 tons
Maximum road speed	51 km/h
Range	250 km
Climb angle (loose sand w/load)	12 degrees
Climb angle (loose sand w/o load)	24 degrees
Fuel consumption (road) approx.	100 l/100 km
Fuel consumption (cross-country) up to	90-100 l/hr
Tow strength on the tow coupling	8000 kg
Tow strength of winch in simple tow	5000 kg

Weights

Gross weight of vehicle	14700 kg
Empty weight (with equipment and fuel)	12150 kg
Cargo load	2550 kg
Ground pressure of front wheels	2700 kg
Ground pressure of tracks	12000 kg
Specific ground pressure on solid surface	4.6 kg/cm^2
Specific ground pressure bound	0.7 kg/cm^2

Dimensions

Total length	7350 mm
Total width	2500 mm
Total height	2770 mm
Height minus canvas cover	2600 mm
Track width of front wheels	2010 mm
Track width of tracks	1900 mm
Ground clearance beneath front axle	430 mm
Ground clearance beneath crosspiece	400 mm
Fording depth	630 mm

Engine

Type	Maybach HL 85 TUKRM
Performance at n= 2600 r.p.m.	185 hp
r.p.m. normal	2500
Number of cylinders	12
Bore and stroke	95 x 100 mm
Cylinder capacity (volume)	8520 cc
Operation	four-stroke
Compression ratio	1:6.5
Lubrication	pressurized by gear-type oil pump
Ignition	magnetic
Ignition operation	automatic
Carburetor	2 inverted-type
Engine cooling system	water circulation (circulation pump)
Oil cooling system	water cooled oil cooler
Valve action: intake and exhaust valve	0.25 mm

Tracks

Type	Zgw. 50/400/200
Width	400 mm
Separation	200 mm
Links per side	55
Length of single track	11000 mm
Ground contact length	2500 mm
Rubber shoes	Type 501 (LDP)

Front Wheels

Tire size Riesenluftreifen	11.25 - 20 extra
Tire pressure	4.5 ATU

Capacities

Fuel, capacity of fuel tanks	210 + 40 = 250 l
Oil, capacity of engine approx.	15 l
Oil, capacity of gearbox	8 l
Oil, capacity of reduction gearbox	14 l
Oil, capacity of steering gearbox	10 l
Oil, capacity of winch	5.5 l
Oil, capacity of drive sprocket gearbox	each 5 l
Oil, capacity of track	each 5 l
Height of oil level in Delbag filter approx.	15 mm
Height of oil level in air compressor	0.6 l
Water, capacity of entire cooling system	56 l

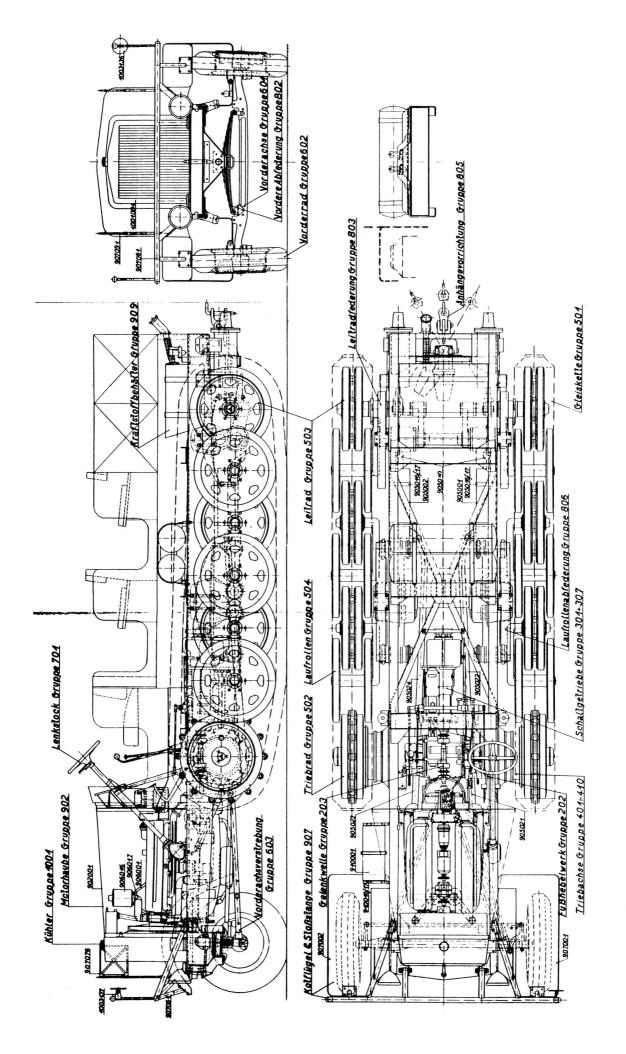

Component sketch for the 12-ton heavy half-tracked prime mover (Sd. Kfz 8), type DBs 7, built from 1934 according to requirement D 608/2 from 1 Feb 1940 (copy). The designation of "heavy all-terrain half-tracked prime mover" was invalidated by this requirement.

This is what the first 12-tonner looked like, taken from D 601 from 8 Nov 1940: "heavy all-terrain half-tracked prime mover (Sd. Kfz. 8). Notice the disc wheels (front) and the shape of the fenders.

Below: 12-ton heavy half-tracked prime mover (Sd. Kfz. 8), type DBs 7 on the training field.

The front axle design is easily recognized on this type DBs 7 vehicle of a driving school. The fundamental design of the axle was only changed on the DB 10.

This and the following photo: 12-tonner DBs 7s towing the 9.2 ton heavy "long 21 cm mortar", which hailed back to the year 1916. These photos were taken in Poland, 1939, by a member of the 7th Infanteriedivision.

This iron-wheeled trailer for the 30.5 cm mortar barrel built by Skoda was photographed in the autumn of 1941 near Smolensk. The tow vehicle was a 12-ton type DBs 7.

Above: the 12-ton heavy half-tracked prime mover (Sd. Kfz. 8), type DBs 8, built from 1938 (chassis number 17101 - 17154) cannot be externally distinguished from the type DB 9 of 1939 (chassis number 440 001 - 440 160), since the only change made was in the type of engine. The regulation D 608/8 (replacement part list issued 9 Oct 1939) treats both vehicles together.

Right: a 12-ton type DBs 8/DB 9 towing a 10.5 cm anti-aircraft gun (with cross mount) on a Sonderanhänger 203 purpose-built trailer.

Left: a 110 cm searchlight dating from the First World War is brought into position by a 12-tonner.

Below: 12-ton heavy half-tracked prime mover (Sd. Kfz. 8), type DBs 8/DB 9.

A 12-tonner with a heavy field howitzer on the training grounds.

Below: the running gear of the 12-ton heavy half-tracked prime mover (Sd. Kfz. 8), type DB 9 (from D 608/7).

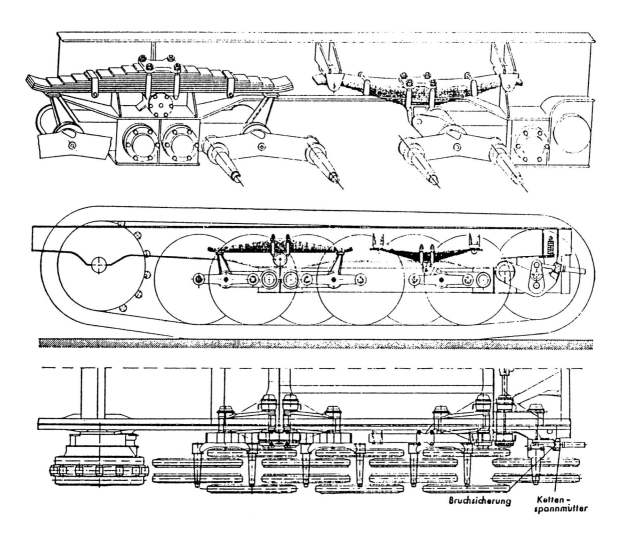

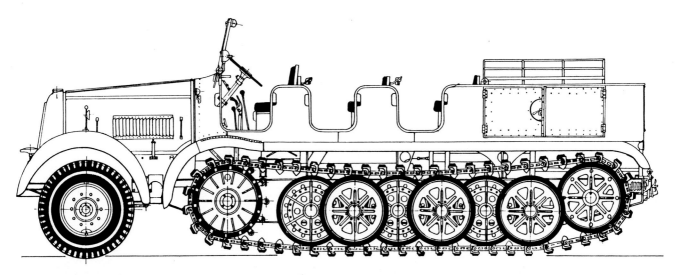

The most important recognition characteristics of the type DB 10 were the front disc wheels and hollow hub (no longer with Mercedes star) of the drive sprocket.

Below: the new torsion bar suspension running gear of the 12-ton heavy half-tracked prime mover (Sd. Kfz. 8), type DB 10 (from D 608/11).

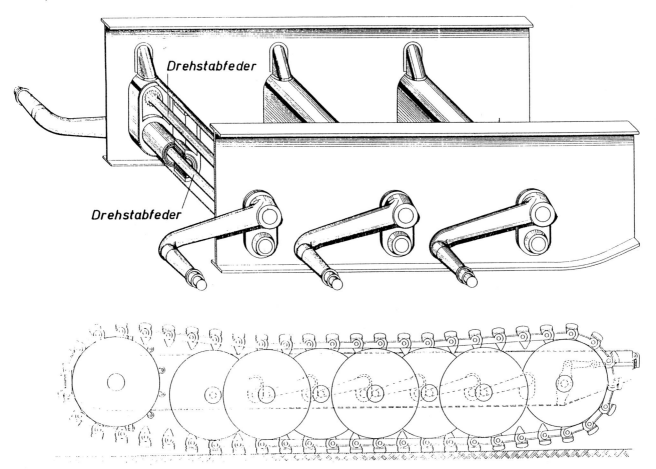

Driving system of a Type DB 10 12-tonner

1. engine coupling, 2. transmission and reduction gearbox, 3. drive sprocket, 4. track drive unit, 5. prime mover brake valve, 6. clutch pedal, 7. brake pedal, 8. gas pedal, 9. hand brake lever, 10. gear shift, 11. reduction gear shift, 12. winch lever, 13. tachograph, 14. radiator shutters operating lever, 15. hand throttle lever, 16. wheel position indicator, 17. signal horn

Factory photo of a 12-ton heavy half-tracked prime mover (Sd. Kfz. 8), type DB 10, built 1939/40 according to D 608/11 from 12 January 1940.

Below: Quite late — for the merciless Russian winter had already descended — D 635/5 appeared on 12 Nov 1941. It was entitled "Vehicles in Winter — Guidelines for Maintenance and Servicing." Here, starting a 12-tonner with the aid of an inertial starter: 1-oil bath air filter, 2-flap for covering superfluous air intake openings during starting, 3-acetylene generator, 4-tarpaulin, 5-inertial starter crank, 6-inertial starter trip lever

12-ton heavy half-tracked prime mover (Sd. Kfz. 8), type DB 10 with an 8.8 cm anti-aircraft gun on a Sonderanhänger(Sd. Ah.) 201 trailer. In this photo the new front axle construction can be seen clearly.

Below: 12-tonner Type DB 10 with an 8.8 cm Flak 36/37 on an Sd. Ah. 202. Each of the three wide benches holds four soldiers.

Rear view of a 12-ton Type DB 10 with an 8.8 cm Flak 36/37.

Below: the 15 cm schwere Feldhaubitze 18 appeared in two versions — the "powered tow" (single piece, hardened rubber tires) and the "horse-drawn tow" (two-piece, steel wheels). This is a photo of a "horse-drawn" version, pulled by a 12-tonner of Artillerieregiment 71 on the road to Stalingrad.

At a road blockade in Rostov: first a 12-tonner with the gun carriage approaches...

...then (below) another 12-tonner with the barrel in tow. This is probably a 2-piece transportable 21 cm Mörser 18. On the left is a standard diesel cargo truck.

12-tonner towing the barrel of a 21 cm Mörser 18.

Below: 12-ton heavy half-tracked prime mover (Sd. Kfz. 8), Type DB 10 with a rather feeble camouflage effort using small trees. It is pulling a barrel trailer.

Above: a 12-tonner being used as a prime mover for a Russian 15.2 cm howitzer.

Below: rear view of a 12-tonner with the carriage of a 21 cm Mörser 18 (may also be a 17 cm Kanone 18 on a mortar carriage).

12-ton heavy half-tracked prime mover (Sd. Kfz. 8), Type DB 10 in the Redya Valley south of Lake Ilmen.

Below: The heavy prime movers were critical to the tank maintenance companies for recovery and transport operations. This photo shows a 12-ton heavy half-tracked prime mover (Sd. Kfz. 8), Type DB 10 of the 9th Panzerdivision transporting a Panzerkampfwagen 38(t) on a lowboy trailer designed for transporting armored vehicles (Sd. Ah. 116).

A Mercedes-Benz Type LG 3000 cargo truck on an armored vehicle lowboy trailer (Sd. Kfz. 115), being towed by a 12-ton heavy half-tracked prime mover (Sd. Kfz. 8), Type DB 10.

Caucasus, autumn 1942: a 12-ton is pulling a 5-tonner (3.7 cm anti-aircraft gun on a 5-ton half-tracked prime mover chassis (Sd. Kfz. 6/2)), providing the reader with an interesting comparison of sizes.

Rear view of 12-ton heavy half-tracked prime movers (Sd. Kfz. 8), Type DB 10 with its crew wearing warm, albeit unusual clothing.

Below: front view of a 12-ton heavy half-tracked prime mover (Sd. Kfz. 8), Type DB 10, minus its military license plate.

12-ton heavy half-tracked prime movers (Sd. Kfz. 8), Type DB 10 during one of the numerous exercises for the planned landings in England.

A 12-tonner in Hungary, 1944, apparently sprayed in "dunkelgelb" (dark yellow) color.

In the Russian mud the half-tracked prime movers were an indispensable aid in maintaining the flow of traffic — even if at only a minimum level. This photo shows a 12-tonner towing a Personenkraftwagen in the Crimea.

Evidenced by their happy faces, these soldiers of the Empire are enjoying their ride in a spacious 12-tonner, captured in North Africa.

Below: 8 May 1945, Tischnowitz (northwest of Brünn): the Wehrmacht is on the retreat. This photo shows a Hetzer being pulled by a 12-tonner.

American soldiers inspect a 12-tonner destroyed during the Battle of the Bulge.

Below: 12-tonners destroyed in Soir, 3 September 1944. Notice the shape of the idler wheel.

A 12-tonner from the Schwere Artillerieabteilung 607 stuck in a ditch.

Below: detail photograph of a 12-ton heavy half-tracked prime mover (Sd. Kfz. 8), Type DB 10

Left: this 12-ton heavy half-tracked prime mover (Sd. Kfz. 8), Type DB 10, has fallen through a small bridge, enabling a top view of the vehicle.

Below: camouflaged 8.8 Flak (Sf.) on a 12-ton heavy half-tracked prime mover (Sd. Kfz. 8) in France, 1940.

Designed for engaging heavily fortified ground targets, particularly the bunkers along the Maginot Line, an armored version of a 12-ton heavy half-tracked prime mover (Sd. Kfz. 8), Type DB 9, was outfitted with a special 8.8 cm Flak. Altogether 25 vehicles were built and were primarily used by the Schwere Panzerjägerabteilung 8. Maximum armor penetrating power was 15 mm. The gun deviated considerably from the standard 8.8 cm Flak and was also carried on a specialized trailer, pulled by an Sd. Kfz. 7.

Above and below: with a height of 2.8 meters, a 12-tonner with an 8.8 cm Flak was an obvious target not to be overlooked on the battlefield. Here are two destroyed self-propelled guns in France, 1940. It is possible that the two photos are of the same vehicle.

THE Sd. Kfz. 9 18-TON HEAVY HALF-TRACKED PRIME MOVER

The 18-tonner was the heaviest half-tracked prime mover produced for the Wehrmacht and was made available to front-line units beginning in 1938. Development of this massive vehicle was undertaken by the company of "Fahrzeug- und Motorenbau GmbH (FAMO) in Breslau starting in 1935. In particular, the 18-tonner was planned for towing the following: lowboy trailers up to 35 tons gross weight, the 24 cm Kanone 3 (towed in six separate components), the 35.5 cm Haubitze M.1 (towed in seven components), the 12.8 cm Flak 40 (on an Sd. Ah. 220 trailer) and other heavy guns.

The first models (designated FM gr. 1) were built by FAMO in 1936 and 1937. In 1938 FAMO introduced the Type F2, little changed from its predecessor and even having the same engine — the Maybach HL 98 TUK. The final version, the Type F3, was built from 1939 to 1944 by FAMO, Vomag and TATRA; this model was fitted with the Maybach HL 108 TUKRM engine.

Unlike other German half-tracks, the basic shape of the 18-tonner changed little throughout its developmental life. About 2500 18-tonners were built. In addition to its above-mentioned tasks of pulling artillery pieces it was also operated by armored maintenance and repair units as a tow and recovery vehicle.

TECHNICAL DATA BASED ON D 671/1 (11 Jan 39) FOR THE 18-TON HEAVY HALF-TRACKED PRIME MOVER (Sd. Kfz. 9), TYPE F 2 (BAUJAHR 1938) and TYPE F 3 (BAUJAHR 1939):

Vehicle Performance

Towed load (weight) normal	18 tons
Maximum road speed	50 km/h
Range	250 km
Climb angle (loose sand w/load)	12 degrees
Climb angle (loose sand w/o load)	24 degrees
Fuel consumption (road) approx.	120 l/100 km
Fuel consumption (cross-country)	up to 100 l/hr
Tow strength of winch in simple tow	5000 kg

Weights

Gross weight of vehicle	18000 kg
Empty weight (with equipment and fuel)	12130 kg
Cargo load (8 men and 1900 kg)	2870 kg
Weight on front wheels	2700 kg
Weight on tracks	12000 kg
Specific ground pressure on solid surface	4.5 kg/cm^2
Specific ground pressure bound	0.7 kg/cm^2

Dimensions

Total length	8250 mm
Total width	2600 mm
Total height	2850 mm
Height minus canvas cover with windscreen	2770 mm
Track width of front wheels	2100 mm
Track width of tracks	2000 mm
Camber of front wheels	2 degrees
front wheel toe in	6 mm
Ground clearance beneath front axle	550 mm
Ground clearance beneath longitudinal frame member	400 mm
Fording depth	800 mm

Engine

	Type F 2	Type F 3
Type	Maybach HL 98 TUK	Maybach HL 108 TUKRM
Performance at n= 2600 r.p.m.	230 hp	
R.P.M. normal	2600	
Number of cylinders	12	
Bore and stroke	95 x 115 mm	100 x 115 mm
Cylinder capacity (volume)	9800 cc	10830 cc
Operation	four-stroke	
Compression ratio	1:6.7	
Lubrication	gear-type oil pump	
Ignition	magnetic	
Ignition operation	automatic	
Carburetor	inverted type	
Engine cooling system	water circulation (circulation pump)	
Oil cooling system	water cooled oil cooler	

Tracks

Type	Zgw. 50/400/200
Width	440 mm
Separation	260 mm
Links per side	47
Length of single track	12220 mm
Ground contact length	2860 mm
Rubber shoes	110 x 240 mm (W 601)

Front Wheels

Tire size	12.75 - 20 extra
Tire pressure	4 ATU

Capacities

Fuel, capacity of fuel tanks	230 + 60 = 290 l
Oil, capacity of engine approx.	18 -20 l
Oil, capacity of gearbox	10 l
Oil, capacity of reduction gearbox	21 l
Oil, capacity of steering gearbox	16 l
Oil, capacity of drive sprocket gearbox	each 3 l
Height of oil level in Delbag filter approx.	35 mm
Water, capacity of entire cooling system	56 l

An 18-ton heavy half-tracked prime mover (Sd. Kfz. 9) as it appeared in D 671/1 from 11 January 1939.

Right: Dashboard of the 18-ton heavy half-tracked prime mover (Sd. Kfz. 9) according to D 671/1 from 11 January 1939:
1. tachometer, 2. hand throttle lever, 3. starter lever, 4. radiator vent control lever, 5. gear shift diagram, 6. starter plate, 7. switch box, 8. rocker switch, 9. oil pressure gauge, 10. twin air brake pressure gauges, 11. rpm gauge, 12. indicator lamp for blue high beam lights, 13. windshield wiper fuse, 14. operating plate for winch, 15. ignition switch, 16. operating plate for centrifugal starter, 17. fuel tank gauge, 18. signal switch, 19. choke switch for horn, 20. temperature gauge, 21. fuse, 22. auxiliary fuel tank gauge, 23. clutch pedal, 24. driving brake pedal, 25. gas pedal, 26. hand brake lever, 27. gear shift lever, 28. wheel position indicator, 29. dimmer switch, 30. signal knob

Tire advertisement from 1943.

Below: 18-ton heavy half-tracked prime mover (Sd. Kfz. 9) towing a Sturmgeschütz on an Sd. Ah. 116 armored transporter. (Welle Archives)

With a height of 2.85 m, a width of 2.6 m and a length of 8.25 m the 18-tonner was a truly interesting and unique half-track. (Welle Archives)

Below: rail transport of an 18-tonner from the repair platoon of Panzerregiment "Grossdeutschland."

Maintenance work on the running gear of an 18-tonner from Panzerregiment "Grossdeutschland."

Below: plowing through the mud in the Volkhov district: an 18-ton heavy half-tracked prime mover (Sd. Kfz. 9) of an engineering unit.

Left: an 18-tonner of the Luftwaffe being shipped to Africa, 1943.

Right: an 18-ton heavy half-tracked prime mover (Sd. Kfz 9) being driven into a heavy amphibious trailer of the "land-wasser-schlepper", or land-water tractor tug.(prototype model).

Left: an 18-tonner in the southern sector of the Eastern Front. On the left is a broken-down "Stalin Organ" with its 132 mm rockets still on their rails.

Left: an 18-ton heavy half-tracked prime mover (Sd. Kfz. 9) being used to tow a gun carriage.

Right: transporting the upper carriage of a "Karl" mortar (60 cm Mörser) of the Schwere Artillerieabteilung 833 on a four-axle Culemeyer trailer near Sevastopol. 18-tonners belonging to this unit wore a "pomegranate" symbol.

Left: an 18-tonner of Schwere Artillerieabteilung 833.

Right: 18-tonners with heavy, two-piece transportable guns. Unfortunately, no further details have been provided about this photograph.

Left: an 18-tonner with barrel trailer and winter camouflage paint in the vicinity of Staraya-Russa.

Right: "WH-512 907" also an 18-ton heavy half-tracked prime mover (Sd. Kfz. 9) of Schwere Artillerieabteilung 833.

When towing a stalled cargo truck, the driver of an 18-tonner had to exercise a considerable amount of careful footwork to avoid pulling the broken-down vehicle apart. "WH-387 983" carries the double-H (Hermann Hoth) designation of Panzergruppe 3, while the division symbol resembles that of the 3rd Infanteriedivision. Notice the unnecessarily masked headlights.

Bavaria 1945. The war is over. An 18-tonner, fully loaded with emaciated soldiers, approaches the Americans.

Below: an 18-ton heavy half-tracked prime mover (Sd. Kfz. 9) tows a night-camouflaged He 11 of 4/KG 26 (Löwengeschwader, 1H + CM) through heavy snow.

From an album of Artillerieregiment 85 of the 101st (leichte) Infanteriedivision: recovering a Sturmgeschütz in Yugoslavia (possibly Pettau) using an 18-tonner, April 1941.

Below: an 18-tonner recovering a damaged Pz, Kpfw. Ausf. J following the tank battle at Sasov in the summer of 1941. On the left is a column of Artillerieregiment 71 ("The Lucky Division", wiped out at Stalingrad).

A picture with a story behind it: during renovation of a Kettenkrad in the early 60s, a collector in the former GDR found a negative strip hidden in the filth and oil of a toolbox. This is one of the pictures from that strip, showing a Sturmgeschütz on an Sd. Ah. 116 being towed by two 18-tonners. Notice the thin tow cable — normally heavy prime movers were coupled together using tow bars.

Below: the propaganda company's original text read: "Budapest, tough fighting. While the din of battle still echoes from the house walls, supply columns roll unimpeded through the streets to the front lines." The photo shows a "Hummel" 15 cm schwere Panzerhaubitze being towed by an 18-tonner — be it for reasons of fuel conservation or because of damage to the vehicle. (Bundesarchiv)

Right: this 18-tonner captured by the Americans shows cut-back fenders, similar to later models. The rounded shape of the fender and the blending of the fender into the track cover nevertheless indicates either an earlier version which has been improved by the front-line troops or an interim model.

Left: this late-model 18-ton heavy half-tracked prime mover (Sd. Kfz. 9) was destroyed in France in 1944.

Right: Italy, 1944: general overhaul of an 18-tonner (late-model) of Panzerjägerabteilung 653.

The late model 18-ton heavy half-tracked prime mover (Sd. Kfz 9) can be recognized chiefly by its simplified front end, with tube bumper and cut-back fenders.

Below: Vehicles of the 3rd Panzergrenadier-Division in Italy, 1943/44. The vehicle in the foreground is a late-model 18-ton heavy half-tracked prime mover (Sd. Kfz. 9); on the right is a Renault 3t AHN cargo truck; a mobile crane, probably a Büssing-NAG 4.5-ton truck with 3-ton traversible crane, can just be made out beneath the olive trees on the left.

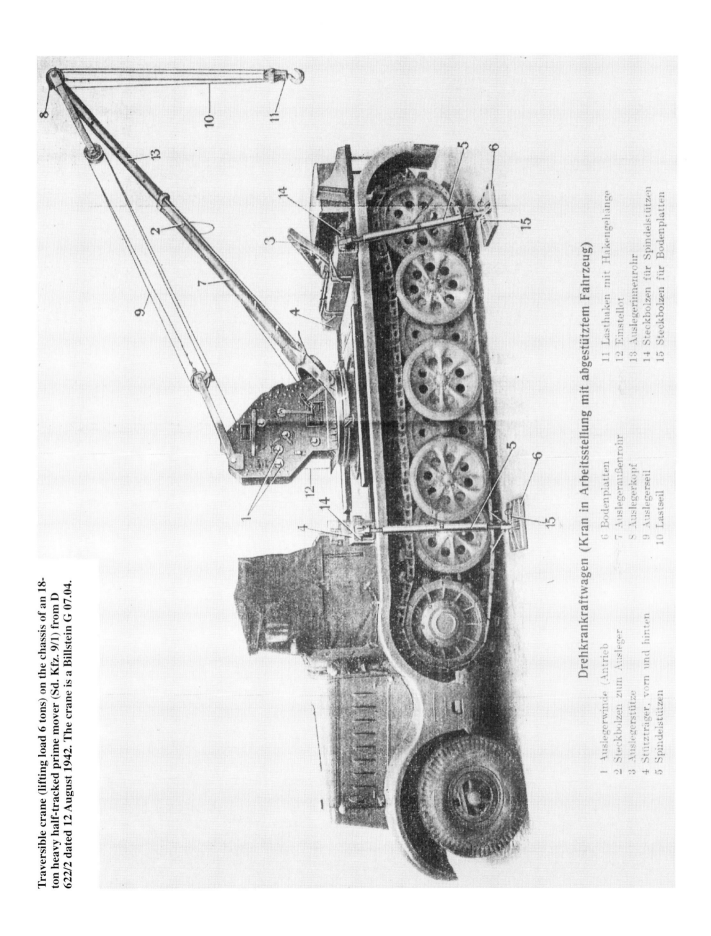

Traversible crane (lifting load 6 tons) on the chassis of an 18-ton heavy half-tracked prime mover (Sd. Kfz. 9/1) from D 622/2 dated 12 August 1942. The crane is a Billstein G 07.04.

Drehkrankraftwagen (Kran in Arbeitsstellung mit abgestütztem Fahrzeug)

1 Auslegerwinde (Antrieb)
2 Steckbolzen zum Ausleger
3 Auslegerstütze
4 Stützträger, vorn und hinten
5 Spindelstützen
6 Bodenplatten
7 Auslegeraußenrohr
8 Auslegerkopf
9 Auslegerseil
10 Lastseil
11 Lasthaken mit Hakengehänge
12 Einstellot
13 Auslegerinnenrohr
14 Steckbolzen für Spindelstützen
15 Steckbolzen für Bodenplatten

The "repair bulls" of the maintenance and recovery units recovered millions' worth of war material, often under the enemy's very noses. The recognition they received for their heroic efforts was severely lacking, despite the greatest sacrifices of life and limb made under combat conditions. The photo shows an Sd. Kfz. 9/1 and an Sd. Kfz. 9 towing a broken-down Tiger tank. (Bundesarchiv)

The winch of an 18-tonner had a pull strength of 7.0 tons. Various spades were tested to provide the corresponding resistance.

Left: the traversible crane with a lift load of 10 tons mounted on an 18-ton heavy prime mover chassis was given the designation Sd. Kfz. 9/2. There are no known photos of this vehicle in operational conditions.

Below: another specialized design was the Flak 8.8 cm on an armored 18-tonner. Only a few were manufactured. Operational photos of this vehicle are not known to exist, either.

The Spielberger German Armor & Military Vehicles Series

Four volumes in the classic series by reknowned German panzer historian Walter Spielberger are now available in new English editions. Known for his emphasis on detail, Spielberger chronicles each production variation and later modifications. Line drawings by Hilary Doyle complement the discussion of each model type.

Size: 8 1/2" x 11"   288 pages   hard cover
over 460 photographs
ISBN: 0-88740-397-2 $39.95

Size: 8 1/2" x 11"   256 pages hard cover
over 240 photographs
ISBN: 0-88740-398-0 $39.95

Size: 8 1/2" x 11"   168 pages hard cover
over 200 photographs
ISBN: 0-88740-448-0 $29.95

Size: 8 1/2" x 11"   168 pages hard cover
over 200 photographs
ISBN: 0-88740-515-0 $29.95